International Association of Fire Chiefs

National Fire Protection Association

S0-ABV-719

Fundamentals of Fire Fighter Skills

THIRD EDITION

JONES & BARTLETT
LEARNING

Jones & Bartlett Learning
World Headquarters
5 Wall Street
Burlington, MA 01803
978-443-5000
info@jblearning.com
www.jblearning.com

National Fire Protection Association
1 Batterymarch Park
Quincy, MA 02169-7471
www.NFPA.org

International Association of Fire Chiefs
4025 Fair Ridge Drive
Fairfax, VA 22033
www.IAFC.org

Jones & Bartlett Learning books and products are available through most bookstores and online booksellers. To contact Jones & Bartlett Learning directly, call 800-832-0034, fax 978-443-8000, or visit our website, www.jblearning.com.

Production Credits

Chief Executive Officer: Ty Field
President: James Homer
SVP, Editor-in-Chief: Michael Johnson
SVP, Chief Marketing Officer: Alison M. Pendergast
Executive Publisher: Kimberly Brophy
VP, Manufacturing and Inventory Control: Therese Connell
VP of Sales, Public Safety Group: Matthew Maniscalco
Director of Sales, Public Safety Group: Patricia Einstein
Executive Acquisitions Editor: Bill Larkin

Senior Editor: Jennifer Deforge-Kling
Production Manager: Jenny L. Corriveau
Senior Marketing Manager: Brian Rooney
Composition: diacriTech
Cover Design: Anne Spencer
Rights & Photo Research Associate: Lian Bruno
Cover Image: Maryland Fire and Rescue Institute
Printing and Binding: Courier Companies
Cover Printing: Courier Companies

To order this product, use ISBN: 978-1-284-04713-4

Library of Congress Cataloging-in-Publication Data
Schottke, David.
  Fundamentals of fire fighter skills / David Schottke, National Fire Protection Association, International Association of Fire Chiefs.—Third edition.
     pages cm
  ISBN 978-1-4496-4152-8 (pbk.)
1. Fire prevention—Handbooks, manuals, etc. 2. Fire extinction—Handbooks, manuals, etc. I. National Fire Protection Association. II. International Association of Fire Chiefs. III. Title.
  TH9151.S276 2012
  628.9'2—dc23
                          2012032314
6048
Printed in the United States of America
16 15 14 13   10 9 8 7 6 5 4 3 2 1

Brief Contents

CHAPTER 1 The Orientation and History of the Fire Service. 2

CHAPTER 2 Fire Fighter Safety 24

CHAPTER 3 Personal Protective Equipment and Self-Contained Breathing Apparatus 42

CHAPTER 4 Fire Service Communications . . . 90

CHAPTER 5 Incident Command System 116

CHAPTER 6 Fire Behavior. 140

CHAPTER 7 Building Construction 166

CHAPTER 8 Portable Fire Extinguishers. . . . 196

CHAPTER 9 Fire Fighter Tools and Equipment. 234

CHAPTER 10 Ropes and Knots. 258

CHAPTER 11 Response and Size-Up. 298

CHAPTER 12 Forcible Entry 318

CHAPTER 13 Ladders 354

CHAPTER 14 Search and Rescue. 402

CHAPTER 15 Ventilation. 440

CHAPTER 16 Water Supply 486

CHAPTER 17 Fire Attack and Foam 550

CHAPTER 18 Fire Fighter Survival. 584

CHAPTER 19 Salvage and Overhaul 610

CHAPTER 20 Fire Fighter Rehabilitation. 642

CHAPTER 21 Wildland and Ground Fires 660

CHAPTER 22 Fire Suppression. 678

CHAPTER 23 Preincident Planning 706

CHAPTER 24 Fire and Emergency Medical Care. . . . . not included

CHAPTER 25 Emergency Medical Care . . . . not included

CHAPTER 26 Vehicle Rescue and Extrication. 808

CHAPTER 27 Assisting Special Rescue Teams. 838

CHAPTER 28 Hazardous Materials: Overview 866

CHAPTER 29 Hazardous Materials: Properties and Effects 878

CHAPTER 30 Hazardous Materials: Recognizing and Identifying the Hazards 900

CHAPTER 31 Hazardous Materials: Implementing a Response. 922

Contents

CHAPTER 1 The Orientation and History of the Fire Service 2

Introduction 4
Fire Fighter Guidelines 4
Fire Fighter Qualifications 5
    Age Requirements 5
    Education Requirements 5
    Medical Requirements 5
    Physical Fitness Requirements 5
    Emergency Medical Care
      Requirements 5
Roles and Responsibilities of
    Fire Fighter I and Fire Fighter II 5
    Roles and Responsibilities for
      Fire Fighter I 5
    Roles and Responsibilities for
      Fire Fighter II 6
Roles Within the Fire Department 6
    General Roles 6
    Specialized Response Roles 7
Working with Other Organizations 7
Fire Department Governance 8
The Organization of the
    Fire Service 8
    Company Types 8
    Other Views of Fire Service
      Organization 10
    Chain of Command 11
    Source of Authority 11
    Basic Principles of
      Organization 12
The History of the Fire Service 13
    The American Fire Service 13
    Building Codes 13
    Training and Education 14
    Fire Equipment 16
    Communications 16
    Paying for Fire Service 17
Fire Service in the United States
    Today 18

CHAPTER 2 Fire Fighter Safety 24

Introduction 26
Causes of Fire Fighter Deaths
    and Injuries 26
Injury Prevention 27
    Standards and Procedures 27
    Personnel 27
    Training 28
    Equipment 28
    Reducing Fire Fighter Injuries
      and Deaths 28
Safety and Health 28
    Employee Assistance Programs 30
Safety During Training 30
Safety During the Emergency
    Response 30
Safe Driving Practices 31
    Laws and Regulations
      Governing Emergency Vehicle
      Operation 31
    Standard Operating Procedures for
      Personal Vehicles 32
    Safe Driving Begins with You 32
    Vehicle Collision Prevention 32
    The Importance of Vehicle
      Maintenance 33
Safety at Emergency Incidents 33
    Teamwork 35
    Accountability 35
    Incident Scene Hazards 36
    Using Tools and Equipment Safely . . . 36
    Electrical Safety 36
    Lifting and Moving 36
    Working in Adverse Weather
      Conditions 36
    Rehabilitation 36
    Violence at the Scene 38
    Mental Well-being 38
Safety at the Fire Station 38
Safety Outside Your Workplace 38

CHAPTER 3   Personal Protective Equipment and Self-Contained Breathing Apparatus **42**

Introduction 44
Personal Protective Equipment 44
  Structural Firefighting
    Ensemble 44
  Donning Personal Protective
    Clothing 51
  Doffing Personal Protective
    Clothing 51
  Care of Personal Protective
    Clothing 51
  Specialized Protective Clothing 54
Respiratory Protection 54
  Respiratory Hazards of Fires 54
  Other Toxic Environments 56
  Conditions That Require
    Respiratory Protection 56
  Types of Breathing Apparatus 56
  SCBA Standards and Regulations 57
  Uses and Limitations of
    Self-Contained Breathing
    Apparatus 58
  Components of Self-Contained
    Breathing Apparatus 59
  Pathway of Air Through
    an SCBA 62
  Skip-Breathing Technique 62
  Mounting Breathing Apparatus 62
  Donning Self-Contained
    Breathing Apparatus 64
  Donning the Face Piece 66
  Safety Precautions for Self-Contained
    Breathing Apparatus 70
  SCBA Use During Emergency
    Operations 70
  Doffing Self-Contained Breathing
    Apparatus 73
  Putting It All Together: Donning the
    Entire PPE Ensemble 73

  SCBA Inspection and
    Maintenance 73
Inspection of SCBA 73
  Operational Testing 75
  Servicing SCBA Cylinders 75
  Replacing SCBA Cylinders 75
  Replacing an SCBA Cylinder on
    Another Fire Fighter 77
  Refilling SCBA Cylinders 78
  Cleaning and Sanitizing SCBA 79

CHAPTER 4   Fire Service Communications **90**

Introduction 92
The Communications Center 92
  Telecommunicators 93
  Communications Facility
    Requirements 93
  Communications Center
    Equipment 94
  Computer-Aided Dispatch 94
  Voice Recorders and Activity Logs . . . 94
  Call Response and Dispatch 95
Communications Center Operations 95
  Receiving and Dispatching
    Emergency Calls 96
  Call Receipt 96
  Location Validation 98
  Call Classification and Prioritization 100
  Unit Selection 100
  Dispatch 100
  Operational Support and
    Coordination 102
  Status Tracking and Deployment
    Management 102
  Touring the Communications
    Center 103
Radio Systems .103
  Radio Equipment 103
  Radio Operation 104
  Using a Radio 106
Records and Reporting 109

Obtaining the Necessary
Information 110
Required Coding Procedures 110
Consequences of Incomplete and
Inaccurate Reports. 110
Using the NFIRS Data Entry Tool. . . 110
Taking Calls: Emergency,
Nonemergency, and
Personal Calls 111

CHAPTER 5 Incident Command System 116

Introduction 118
History of the Incident
Command System 118
Characteristics of the
Incident Command System 119
Jurisdictional Authority 119
All-Risk, All-Hazard System 120
Everyday Applicability 121
Unity of Command 121
Span of Control 121
Modular Organization 121
Common Terminology 122
Integrated Communications 122
Consolidated Incident Action Plans . 122
Designated Incident Facilities 122
Resource Management 122
The ICS Organization 122
Command 123
General Staff Functions 124
Standard ICS Concepts and
Terminology 126
Single Resources and Crews 126
Branches 127
Location Designators 127
Implementing ICS 128
Standard Position Titles 130
Working Within the Incident Command
System 130
Responsibilities of the First-Arriving
Fire Fighters 130
Confirming the Command 131

CHAPTER 6 Fire Behavior 140

Introduction 142
The Chemistry of Fire 142
What Is Fire? 142
States of Matter 142
Fuels 143
Types of Energy 143
Electrical Energy 143
Conservation of Energy 144
Conditions Needed for Fire 144
Chemistry of Combustion 144
Products of Combustion 145
Fire Spread 146
Methods of Extinguishment 148
Classes of Fire 148
Class A Fires 148
Class B Fires 148
Class C Fires 149
Class D Fires 149
Class K Fires 149
Characteristics of
Solid-Fuel Fires 149
Solid-Fuel Fire Development 150
Characteristics of a Room-and-
Contents Fire 152
Special Considerations 154
Fire Behavior in Modern
Structures 156
Wind Effect 156
Characteristics of
Liquid-Fuel Fires 157
Characteristics of Gas-Fuel Fires 157
Vapor Density 157
Flammability Limits 158
Boiling Liquid/Expanding Vapor
Explosions 158
Smoke Reading 158
Step 1: Determine the Key
Attributes of Smoke 159
Step 2: Determine What Is Influencing
the Key Attributes 160

Step 3: Determine the Rate of
Change 160
Step 4: Predict the Event 160
Smoke Reading Through a Door . . . 161

CHAPTER 7

Building Construction. 166

Introduction 168
Occupancy. 168
Contents 168
Types of Construction Materials 169
Masonry. 169
Concrete 169
Steel. 170
Other Metals 171
Glass 171
Gypsum Board. 171
Wood. 172
Plastics. 173
Types of Construction 173
Type I Construction:
Fire Resistive 174
Type II Construction:
Noncombustible 175
Type III Construction: Ordinary. . . . 176
Type IV Construction:
Heavy Timber 176
Type V Construction:
Wood Frame 177
Building Components. 180
Foundations. 180
Floors and Ceilings 180
Roofs 181
Trusses. 184
Walls 186
Doors and Windows 188
Interior Finishes and Floor
Coverings 189
Manufactured Housing 189
Buildings Under Construction
or Demolition 190
Preincident Planning and
Incident Size-Up 190

CHAPTER 8

Portable Fire Extinguishers 196

Introduction 198
Purposes of Fire Extinguishers 199
Use of Portable Fire Extinguishers in
Incipient-Stage Fires 199
Special Extinguishing Agents 199
Classes of Fires. 200
Class A Fires 200
Class B Fires. 200
Class C Fires 201
Class D Fires 201
Class K Fires 201
Classification of Fire Extinguishers. . . 202
Labeling of Fire Extinguishers 202
Traditional Lettering System. 202
Pictograph Labeling System 203
Fire Extinguisher Placement 203
Classifying Area Hazards 204
Determining the Most Appropriate
Placement of Fire
Extinguishers 205
Methods of Fire Extinguishment 205
Cooling the Fuel 206
Cutting Off the Supply of
Oxygen 206
Interrupting the Chain of
Reactions. 206
Types of Extinguishing Agents 206
Water. 206
Dry Chemicals 206
Carbon Dioxide 207
Foam 208
Wet Chemicals 208
Halogenated Agents 209
Dry Powder 209
Fire Extinguisher Design 209
Portable Fire Extinguisher
Components 209
Wheeled Fire Extinguishers 211
Fire Extinguisher Characteristics. 211
Water Extinguishers. 211

Dry-Chemical Extinguishers 212
Carbon Dioxide Extinguishers 214
Class B Foam Extinguishers 215
Wet-Chemical Extinguishers 215
Halogenated-Agent Extinguishers . . 215
Dry-Powder Extinguishing Agents . . 215
Use of Fire Extinguishers 216
Locating a Fire Extinguisher. 216
Selecting the Proper Fire
    Extinguisher 216
Transporting a Fire Extinguisher . . . 216
Basic Steps of Fire Extinguisher
    Operation 218
Ensure Your Personal Safety 218
The Care of Fire Extinguishers 225
Inspection 225
Maintenance 225
Recharging 226
Hydrostatic Testing 227

CHAPTER 9

Fire Fighter Tools and Equipment 234

Introduction 236
General Considerations 236
Safety 236
Conditions of Use/Operating
    Conditions 236
Effective Use 236
Functions 237
Rotating Tools 237
Pushing/Pulling Tools 239
Prying/Spreading Tools 240
Striking Tools 242
Cutting Tools 244
Multiple-Function Tools 248
Special-Use Tools 248
Phases of Use 248
Response/Size-Up 248
Forcible Entry 248
Interior Firefighting Tools and
    Equipment 249
Overhaul Tools and Equipment 251

Tool Staging 252
Maintenance 252
Cleaning and Inspecting
    Hand Tools 252
Cleaning and Inspecting Power
    Equipment and Tools 252
Steps for Cleaning and Inspecting
    Power Tools 253

CHAPTER 10

Ropes and Knots 258

Introduction 260
Types of Rope 260
Life Safety Rope 260
Personal Escape Ropes 261
Utility Rope 261
Rope Materials 262
Natural Fibers 262
Synthetic Fibers 262
Rope Construction 263
Twisted and Braided Rope 263
Kernmantle Rope 264
Dynamic and Static Rope 265
Dynamic and Static Kernmantle
    Ropes 265
Rope Strength 265
Tec·ical Rescue Hardware 265
Harnesses 266
Rope Rescue 266
Rope Maintenance 268
Care for the Rope 268
Clean the Rope 268
Inspect the Rope 269
Store the Rope 270
Knot s 271
Terminology 271
Safety Knot 272
Hitches 272
Loop Knots 279
Bends 279
Hoisting 287
Hoisting an Axe 287

Hoisting a Pike Pole 288
Hoisting a Ladder 289
Hoisting a Charged Hose Line . . . 289
Hoisting an Uncharged
    Hose Line 289
Hoisting an Exhaust Fan or
    Power Tool 289

CHAPTER 11   Response and Size-Up 298

Introduction 300
Response 300
Alarm Receipt 301
Riding the Apparatus. 301
Emergency Response. 302
Prohibited Practices. 303
Dismounting a Stopped
    Apparatus 303
Traffic Safety on the Scene 303
Arrival at the Incident Scene. 304
Personnel Accountability
    System 304
Controlling Utilities. 305
Size-Up 306
Managing Information. 306
Resources 310
Incident Action Plan 311
Rescue. 312
Exposure Protection 312
Confinement. 313
Extinguishment. 313
Salvage and Overhaul 313

CHAPTER 12   Forcible Entry 318

Introduction 320
Forcible Entry Situations. 320
Forcible Entry Tools 321
General Tool Safety 321
General Carrying Tips 321
General Maintenance Tips 321
Types of Forcible Entry Tools. . . . 322
Doors 325
Basic Door Construction 325

Construction Materials 326
Types of Doors. 327
Windows. 331
Safety 332
Glass Construction 332
Frame Designs. 333
Locks 338
Parts of a Door Lock 338
Parts of a Padlock 338
Safety 339
Types of Locks. 339
Forcing Entry Through Security
    Gates and Windows 342
Breaching Walls and Floors 344
Load-Bearing/Nonbearing
    Walls 344
Exterior Walls 344
Interior Walls 344
Floors 345
Vehicle Entry. 347
Systematic Forcible Entry 347
Forcible Entry and Salvage 349
Before Entry 349
After Entry 349

CHAPTER 13   Ladders 354

Introduction 356
Functions of a Ladder 356
Secondary Functions of
    Ladders. 357
Ladder Construction 357
Basic Ladder Components. 357
Extension Ladder Components . . 358
Types of Ladders 359
Aerial Apparatus 359
Portable Ladders 359
Inspection, Maintenance, and Service
    Testing of Portable Ladders. . . . 362
Inspection 363
Maintenance 363
Cleaning 364
Service Testing. 365

Ladder Safety. 366
  General Safety Requirements. . . . 366
  Lifting and Moving Ladders. 366
  Placement of Ground Ladders . . . 366
  Working on a Ladder. 367
  Rescue. 368
  Ladder Damage 368
Using Portable Ladders. 368
  Selecting the Ladder 369
  Removing the Ladder from the
    Apparatus. 370
  Lifting Ladders 370
  Carrying Ladders. 371
  Placing a Ladder 379
  Raising a Ladder 381
  Securing the Ladder 388
  Climbing the Ladder 389
  Dismounting the Ladder 391
  Working from a Ladder 393
Inspect a Chimney 396

CHAPTER 14 Search and Rescue. 402

Introduction 404
Search and Rescue Operations . . . 404
  Coordinating Search and
    Rescue Operations with
    Fire Suppression. 404
  Search and Rescue Size-Up 405
  Search Coordination 407
  Search Priorities 407
Primary Search. 407
Search Tec·iques 408
  Search Patterns 409
  Secondary Search 412
Search Safety. 413
  Risk Management 413
  Search and Rescue Equipment. . . 414
  Methods to Determine Whether
    an Area Is Tenable. 414
Rescue Tec·iques 415
  Shelter-in-Place 415
  Exit Assist 416

Simple Victim Carries 416
Emergency Drags. 420
Assisting a Person Down a
  Ground Ladder. 428
Removal of Victims by Ladders . . 435

CHAPTER 15 Ventilation. 440

Introduction 442
Fire Behavior and Ventilation 442
Benefits of Proper Ventilation 444
Backdraft and Flashover
  Considerations 445
  Backdraft. 445
  Flashover. 445
Factors Affecting Ventilation. 445
  Convection 446
  Mechanical Ventilation 446
  Wind and Atmospheric Forces . . 446
Building Construction
  Considerations 447
  Fire-Resistive Construction 447
  Ordinary Construction 448
  Wood-Frame Construction 448
Tactical Priorities 449
  Venting for Life Safety 449
  Venting for Fire Containment . . . 449
  Venting for Property
    Conservation 449
Location and Extent of Smoke and
  Fire Conditions 449
Types of Ventilation. 450
  Horizontal Ventilation 450
  Mechanical Ventilation 455
  Vertical Ventilation 459
Basic Indicators of Roof
  Collapse. 463
Roof Construction 463
  Roof Support Structures 463
  Roof Coverings 463
  Effects of Roof Construction on
    Fire Resistance 463

Connecting Supply Hose Lines
to Standpipe and Sprinkler
Systems 539
Replacing a Defective Section of
Hose 542
Draining and Picking Up Hose . . 543
Unloading Hose 544

CHAPTER 17   Fire Attack and Foam 550

Introduction 552
Attack Hose 552
Sizes of Attack Lines 552
Attack Line Evolutions 553
Preconnected Attack Lines 553
Wyed Lines 559
Advancing Attack Lines 559
Extending an Attack Line 566
Advancing an Attack Line from a
Standpipe Outlet 567
Replacing a Defective Section of
Hose 567
Nozzles 568
Nozzle Shut-Offs 569
Smooth-Bore Nozzles 569
Fog-Stream Nozzles 569
Other Types of Nozzles 570
Nozzle Maintenance and
Inspection 572
Foam 573
Foam Classifications 573
Foam Concentrates 574
Foam Equipment 575
Foam Application 576
Foam Application Techniques . . . 576
Backup Resources 577
Foam Apparatus 578

CHAPTER 18   Fire Fighter Survival 584

Introduction 586
Risk–Benefit Analysis 586
Hazard Indicators 587
Safe Operating Procedures 588

Rules of Engagement for
Fire Fighter Survival 588
Team Integrity 588
Personnel Accountability
System 589
Emergency Communications
Procedures 590
Rapid Intervention
Companies/Crews 591
Fire Fighter Survival Procedures . . 593
Maintaining Orientation 593
Self-Rescue 593
Safe Locations 596
Air Management 600
Rescuing a Downed Fire Fighter 601
Rehabilitation 602
Counseling and Critical
Incident Stress 603

CHAPTER 19   Salvage and Overhaul 610

Introduction 612
Lighting 612
Safety Principles and Practices . . . 612
Lighting Equipment 613
Battery-Powered Lights 613
Electrical Generators 614
Lighting Methods 614
Cleaning and Maintenance 615
Salvage Overview 616
Safety Considerations During
Salvage Operations 617
Salvage Tools 617
Using Salvage Tec·iques to Prevent
Water Damage 617
Deactivating Sprinklers 618
Removing Water 622
Using Salvage Tec·iques to Limit
Smoke and Heat Damage 626
Salvage Covers 626
Salvage Cover Maintenance 628
Floor Runners 631
Other Salvage Operations 631

Overhaul Overview 632
  Safety Considerations During
    Overhaul 633
  Coordinating Overhaul with
    Fire Investigators 633
  Where to Overhaul 634
Overhaul Tec·iques 635
  Overhaul Tools 635
  Opening Walls and Ceilings 636

CHAPTER 20   Fire Fighter Rehabilitation. 642

Introduction 644
Factors, Causes, and Need for
  Rehabilitation 645
  Personal Protective Equipment . . 645
  Dehydration 646
  Energy Consumption 646
  Tolerance for Stress 647
  The Body's Need for
    Rehabilitation 647
Types of Incidents Affecting
  Fire Fighter Rehabilitation 648
  Extended Fire Incidents 648
  Other Types of Incidents Requiring
    Rehabilitation 650
How Does Rehabilitation Work?. . . .651
  Relief from Climatic Conditions . . 652
  Rest and Recovery. 652
  Active or Passive Cooling or
    Warming. 652
  Rehydration and Calorie
    Replacement. 653
  Medical Monitoring. 655
  Member Accountability. 656
  Release and Reassignment 656
Personal Responsibility in
  Rehabilitation 656

CHAPTER 21   Wildland and Ground Fires 660

Introduction 662
Wildland and Ground Fires and the
  Fire Triangle. 662

Fuel. 663
Oxygen 664
Heat 664
Other Factors That Affect Wildland
  Fires 665
  Weather. 665
  Topography. 665
Extinguishing Wildland Fires 666
  Anatomy of a Wildland Fire 666
  Methods of Extinguishment. 666
  Types of Attacks. 667
Priorities of Attack. 669
  Fire Apparatus Used for Wildland
    Fires 669
  Safety in Wildland Firefighting . . 671
  Personal Protective Equipment . . 671
The Challenge of the Wildland–
  Urban Interface. 673
Wildland Fire Safety 674

CHAPTER 22   Fire Suppression. 678

Introduction 680
Offensive Versus Defensive
  Operations. 680
  Command Considerations. 681
Operating Hose Lines 681
  Fire Streams 682
  Interior Fire Attack 682
  Large Handlines 684
  Master Stream Devices. 687
Protecting Exposures 690
Ventilation 690
Specific Fire-Ground
  Operations. 690
  Concealed-Space Fires. 690
  Basement Fires 692
  Fires Above Ground Level 692
  Fires in Large Buildings. 694
  Fires in Buildings During
    Construction, Renovation,
    or Demolition. 694
  Fires in Lumberyards. 694